THE NEXT

LEVEL

OBSERVER

EFFECT

Taking Observer Effect To The Deeper Level Of Application.

HITEN SHELAR

Preface

The observer effect of quantum physics just does not limit itself to quantum physics. Quantum physics gives rise to the reality. In quantum physics, the theory of observer effect was just limited to the collapse of the wave function. The observer collapses the wave function. The wave function gives meaning to reality. So in that sense, the observer gives meaning to reality. This role of observers is not discussed in detail after that in any textbook of science.

Numerous great physicists have gone through and are still going through various sources of science and other than science to learn more about the observers. The person with a real passion for the research and knowledge of our universe studies various sources about the "observer" that gives rise to reality. How the observer gives rise to reality is broken down in detail in this book. The main purpose of this book is to educate people more about the observer, and the observer effect and take the application of the observer effect in various branches like physics, astrophysics, extra-terrestrial life, and in the cosmic perspective.

Contents

1, The Mystery. — 11

2. The Quantum World. — 14

3. The Observer Effect. — 18

4. The Universe Created by chance. — 21

5. The Next Level Application. — 24

Why are we alone in this universe?

6, The Information. — 28

7. Beware of fake scientific media. — 35

8. Science fiction movies and freedom of information. — 43

9. The bizarre way of making . — 46

10. Why are we so special? — 49

11. Why are there no alien observers.   55

The Divine Dilemma.

12, The Information (2).   65

13. The conversation.   69

14. Understanding the truth.   73

15. Symmetry and Non-symmetry of   78
the universe.

16. Good or Bad Nature of God.   83

17. Atheists and Flat Earthers.   90

18. Atheism is not always Bad.   94

19. Creator or created under the   97
influence of?

20. What happens when a high non-symmetry is conducted in the high symmetry universe?   100

21. So does God exist?   102

Above The Cosmic Perspective.

22. The Cosmic Perspective.   107

23. The Hyperspace.   111

24. The idea .   119

25. Do we really don't matter in this universe?   121

26. How do we matter in this universe?   124

27. The Interpretation.   132

28. The Role of Technology.   137

29. We Know Everything. 141

30. Things that terrify humanity to lose their faith. 145

31. The action and the fruit of the action 148

32. The Wrong Knowledge. 154

33. The Inherited Fruit. 158

34. The Money. 163

35. The relationship between two observers. 168

36. Why knowledge of "self" is important? 172

37. The Study of Observers. 174

38. The creation of this universe. 176

39. The evolution of life on the earth. 180

40. The Higher Dimensional Perspective.

ONE

The Mystery.

Humanity always knew, that there is something special about it. No matter how much the world appears classical and gives no significance to humanity, man knew that there is a point in science that connects humans with the creation of the universe itself. And he kept this ideology alive for generations. The idea that a human is special in doesn't seem to make sense. Why not? That's because of the nature of the universe that always chooses the highest symmetry. And due to high symmetry, we can't "directly" relate the creation with us in theory. And thus we may not find ourselves as of significance. Since the symmetric creation supports the creation of a "process", we think that the process is the creator. But the process is created under the cause of observers of this universe. Although high symmetry does not mean "no non-symmetry".   In the world of subatomic particles, where reality gets its meaning, we start seeing something special in us/ humans.

And it was very much expected to find this truth. There is a good number of examples of scientists who at the later end of their career as a scientist turn towards knowing the observers of this universe.

"Science cannot solve the ultimate mystery of nature. And that is because, in the last analysis, we ourselves are a part of the mystery that we are trying to solve."

- Max Planck

So, this is what we come across when we know that the classical world we see around us is created under choice. The choice that is made under the "cause" is discussed ahead in this book.

It can be a common notion to distinguish physics from magic by its nature to surprise us the least. Both physics and magic are but the processes. The process of creating some results. The process we are prone to is what we call "Physics". The process we are not prone to is called "Magic". The process of physics has a high symmetry and the process of magic has less symmetry. This topic is discussed in detail in this book. The magic discussed here is what we see at the tiniest scales.

I would not although like to call it magic. I would just call it the "less symmetric process". Let's now understand the Quantum world, a world where these less symmetrical processes occur.

TWO

The Quantum World.

At the subatomic scale, through our experiments, we find that the sub-atomic particles do not behave the way the world around us does. Because, unlike Newtonian physics where the laws of physics govern the fate of the system, the world of subatomic particles works on a particular choice made, a notion that outruns our understanding of physics. And even today, physics does not know why the quantum world (the world of subatomic particles) behaves this way.

To describe, the quantum world is full of choices. And that choice attains a specific value at any given point throughout the universe. And the physics can't predict this choice. But it says to make a logical guess on the choice to be made. And I don't find a logic in it. Because that logic basically is just guesswork done by looking at the patterns. This is how the subject of physics is made. It has been created by looking at the patterns. We look at the pattern in this universe and if it applies everywhere, we label it "The Law". Physics is nothing but our labeling of the patterns we see and witness. This pattern is so well applicable that it can even help to predict things we have never seen. And where these patterns break, our guesswork fails.

But how is this guesswork of physics so accurate? This is because the patterns we make our laws on apply all over the universe and in every frame. This is due to the symmetry of our universe. Our universe creates everything performing the highest symmetry. Just like expecting an odd answer at the hundred-thousandth position after having identical outcomes throughout is a bad guess. And making a good guess is very easy in such a case. The good guess is called physics. So it's more likely for this good guess to win, due to the same patterns seen all over the universe. But physics can never explain these patterns. As at the end of every "why" in physics, physics has no logic and we simply say "This is how it works". This statement simply means that this is how a choice is made. In this book, a choice maker and the choice made have been discussed.

In the quantum world, the choice made belonged to infinite sample space before the choice is made. But the sample space seen in our universe has a limit. And it is not infinite. After plotting the graph of this sample space to predict the choice in our universe, we see peaks and troughs in the graph. This variation is calculated based on the patterns seen by the choices seen to be made in the past. And then we say that physics predicted this. The advanced level physics is just the network of these learnings from the patterns in the past.

Sometimes we also try to apply these patterns to the patterns we never explored, and many of them turn out correct and sometimes it does not. After trimming the graph of the world of quantum with the patterns of past, what we are left with is the part that shows the non-symmetry of the universe. The same universe where everything is carried out with high symmetry, that's our universe. This graph is called the wave function in quantum mechanics. And by that, it's understood that the patterns that get created in the whole universe, learning to which we make an obvious guess and then call that guess "physics" are based on the quantum world, where a definite choice is made. The choice made on the scale of the Quantum World shows no patterns although. They seem to attain a specific value for some purpose. The same purpose that has intrigued many great scientists in knowing more about our universe.

THREE

The Observer Effect.

The "observers" of this universe have a deep relation with this universe. Whenever our role comes in science, we are entitled as the "observers" of this universe. Because science expects to see the non-symmetric relation between us and the universe. And science has no reason behind this partiality between symmetry and asymmetry. But in the world of quantum, this non-symmetric is seen. Where every choice of the world of choices of quantum world attains a particular value under the cause discussed ahead in this book. In the world of quantum, everything is a choice, and a choice chooser is the observer.

This is the observer effect. In quantum physics, it's said that "The observer collapses the wave function". The wave function in quantum physics is nothing but the sample space of choices shaped by the level of symmetry this universe works. And the choice maker is the observer. Alternatively, it is also said that the measurement collapses the wave function and the one who makes the measurement is the observer.

Now, the observer in 3 dimensions can't be everywhere in the universe to make the choice. So, the correct way to look at this would be to understand the phrase "creating under the cause of". The choice is made under the cause of the observer. Like an egg is older than the life inside it, although the egg is formed under the cause of life inside it. The choice of every aspect of the universe has been made under the cause of observers created. In this book, the true cause of observers is discussed. And to note, the wave function collapses only under this cause.

An egg has a purpose to create a life and it has formed under the same cause or purpose. Similarly, The universe has a purpose as well. Just because the planets and stars are not performing some highly symmetric actions, doesn't mean that the universe is purposeless. The purpose of this universe is discussed later in this book. This book is going to discuss on observer effect to a larger extent of its application ahead.

There are innumerable examples of the things creating themselves under the cause of the observer. For example, who decides what light should behave like as a wave and particle at a particular event? Well, the decider doesn't sit and decide, instead what always happens Is the action or choice-making under the cause of the decider. The main cause of the decider is the purpose of this universe (discussed in this book ahead).

Throughout this book, the word observer has been used. As it's a word that reminds us about science.

FOUR

The Universe Created By Chance.

Some people say that our universe is just a result of chance. Such people have limited scientific knowledge as they don't understand that this change is still "on live". So, every single instant this choice is made all over the universe. This choice decides the fate of the entire universe. As everything gets created from a choice made from the very basic point of the universe. Thus the dice of chance is rolled all the time. It's a live process. And since the first instant of creation of our universe, this dice is just fixed at one value. Do we still call this a chance? How can we? We can't. Out of infinite chances, if a single outcome is repeating itself every single time, then that's not the subject of probability.

Let's take two sides of a coin, heads and tails. And for an outcome to come, there are two elements in the sample space. Let's say we tossed the coin 100000 times and every single time the outcome came out to be 'heads'. In such a case, we can't call it a matter of chance. And for sure we will doubt our coin. Similarly, since the creation of the universe, every single instant has been making the same choice. And we still call that a result of chance? Of course no. The choice always happens under the cause of observers.

Just like by closing the window of a smooth-running airplane, one can say that the airplane itself is not flying. And just like one can call earth as flat by taking the advantage of immense size of the spherical earth. One can call the mystery result of our hypothetical toss as still a matter of chance. Such people are not ready, to accept the reality. They can be compared to the flat earther society.

FIVE

The Next Level Application.

Taking the observer effect to the next level of application is important. The reason why it's important is what one will realize as one reaches the end of this book. The observer is the choice maker. And every choice happens under his cause. This fact cannot be ignored in any aspect. Ignoring the facts can lead us to bizarre and non-real conclusions. This has been one of the themes of this book.

This book compiles the knowledge of various experts gathered throughout the discussion. A great scientist is never institutionalized and never interested in it's respective standards. He just cares about knowing the universe. And the topic of observers and the universe is unavoidable in such a case. And the next application of that topic is given in this book.

WHY ARE
WE
ALONE
IN THIS UNIVERSE
(Why There is No Other intelligent Life in Our Univers

DESTROYING THE DELUSION
SCIENCE MEDIA

WHY ARE WE ALONE IN THIS UNIVERSE?

SIX

The Information.

There are around 8 billion people in the world. And so we have huge diversities in the ideology of the people and the information they believe. There are two types of information, namely fake information and true information. In the diverse human society, there is no difference between fake and true information. Research has shown how tough it is to correct the spread of misinformation among people. This is because very few are interested in seeing another video or content on that topic. For example, if a person's name is wrongly introduced in the media then however one tries to correct it later, it's a much tougher process. And people continue to call the person by the name first introduced. Research has proven this so many times. So it's easy to spread the information at first but very difficult to erase or correct it later once it is spread.

And the one thing that hurts is that wrong information has its own PH.Ds. Such a belief is not only a waste of time but also of the precious life force. For example, the flat earth society. This society is taking absolute benefit of the fact that the world's economy is small enough, to not take the entire humanity to space and actually see whether Earth is flat. To prove a spherical earth as it is, one is not necessarily required to go to space. Let's say you visit two places on Earth 5000 km apart longitudinally. Now, measure the length of a shadow of a wooden stick in both places, and you will get different lengths. In the case of flat earth, no matter where you go, the length of the shadow should be the same. Besides, being proven wrong so many times, the flat earth society still exists.

Often times people believe in a particular information just because they like that information. For example, a kid after watching a sci-fi movie expects fictitious characters or concepts to appear in the real universe. This does not only happen with kids but also with grown-ups. I remember 2 years ago when I met a person who was talking about quantum physics and power of mind and related topics. And here is how our entire conversations go on covering the main parts(not taking the real name of the person in the conversation):

Rahul: Hi, do you believe in the topic of Quantum physics and the power of mind?

Me: Of course! The observers and the universe have a deep relation. And I also know one of the people taking this subject to the next level and communicating to the world, bringing a revolution. One of these people is Dr. Joe Dispenza.

Rahul: (Gets extremely excited) Yes, it's all about it, I loved to see that you know everything about it.

Me: Yes, thanks

Rahul: Yes literally man, classical physics is just nuts, quantum physics is the king, it already has proven that faster than light travel is already possible,

Here in the conversation, everything was ok until the last part where a person became super excited, in fact extremely excited that he was making wrong conclusions. I agreed to whatever he was saying. But what I didn't like is the fact that he was baselessly proving the solid concepts of physics wrong in overexcitement. This is what happens when one is in over excitement.

In quantum physics of course we can get faster than light results, but that's not a travel at all. In the process of quantum tunneling the electron does not travel to the opposite side of the wall, it instead defines itself at the position in space. Similarly in quantum entanglement, there is no travel of any distance. So, being extremely excited, Rahul was telling the concept of physics to be wrong. Now, I could understand it, but what about non-professionals in physics. They will simply make the wrong conclusion and spread misinformation further.

I would like to give one more example here:

Once back in 2018, I visited a religious place, where I met one of the members of that religious society. The member was one of the prime members of that society, who was supposed to spread that religion further. And of course, they have a firm belief in god. But while explaining so, he was spreading the misinformation to the society. Here is how the conversation goes:

Religious Society Member: Hi, do you believe in God?

Me: Of course, I do.

Religious Society Member: What profession you are into?

Me: I love Astrophysics (As a school student back then)

Religious Society Member: Nice, Do you really believe in the Big Bang theory? I mean let's say you have liquid concrete and you bomb it, so will that make a building? He was explaining the process of making of life as too complex a process. And of course, I agree to that. But his attempt to prove the Big Bang theory wrong is just a spreading of wrong information. Our evidence (Cosmic Microwave background) strongly proves the Big Bang theory. In fact, the creator of this universe or God can be explained with the inclusion of the Big Bang theory so well and so it has no controversy with Big Bang at all. However, due to the limited knowledge about physics, the society members were spreading the wrong information about physics. Again, a professional can understand that, but what about general people? They would simply get misinformed and spread the misinformation still further.

I have numerous examples of this where people talk about any wrong shit as a cheap support to their argument. And when the wrong information is spread through media, it becomes tough to erase it from people's minds. This applies not only to science media but to every other media available. Although this book focuses on the scientific media throughout.

Beware of Fake Scientific Media.

Always beware of whom you are listening to. If you are studying astrophysics, you are interested in knowing about our universe and everything, a person then you should always listen to is a Physicist + Philosopher. This universe is always best explained in this approach. Mark my words here, "This universe can never be described by physics alone". And one always needs philosophy to explain the nature of our universe. A physicist + philosopher or a physicist who is also a philosopher does it right. He explains the universe with like 99% of understood physics and 1% of philosophy. A percentage of philosophy that also supports our observation. So philosophy not only explains the origin of physics but also joins the gap where physics fails to explain this universe. A person who says everything is physics lacks knowledge and has a limited understanding of the universe. So many times it happens that philosophy teaches one more of physics and physics strengthens the philosophy. The philosophy of a physicist is never bewildered, it always makes claims without any violation of the law of physics and the observations we make. In fact, it supports the physics. One should always listen to such individuals. This book answers the question "Why we are alone in this universe". We have millions of star systems in our own galaxy and there are billions and trillions of galaxies in the observable universe. A physicist will say of course,

I mean of course! Aliens do exist. And if he is a science fiction fan, he will increase his ignorance of the true knowledge further. In this book, I will explain to you how are we alone in this universe. And I will do it with Physics and Philosophy. The explanation will contain more of a physics part. In fact, most physicists will believe that it's only the physics which I am talking about.

There are of course the science communicators who are only physicists and not philosophers, who don't make stupid claims. They just make their audience feel the beauty of physics. And they just don't talk about things that require him to become a physicist + philosopher. And I respect such communicators since they don't spread misinformation.

I personally do watch a lot of Astrophysics content on Social Media. I have seen reels disappearing after their claims are proven wrong. To mention the content, I am talking about, a 2022-2023 reel that went viral on Instagram and various social media platforms. The reel was about the picture of spacetime curvature. Look at the image below, this is how we imagine the concept of spacetime curvature caused by some mass.

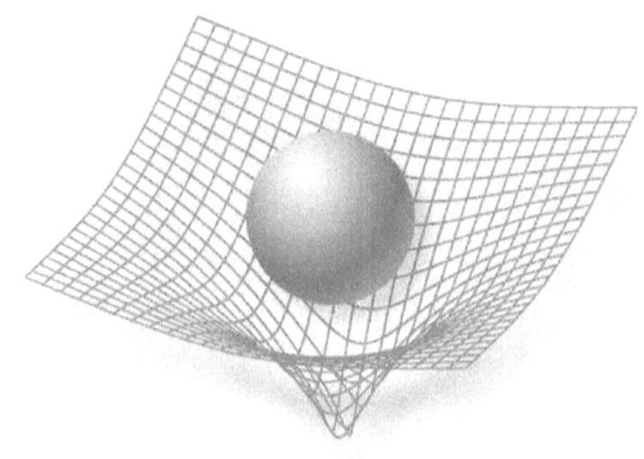

Figure (a)

And now a viral reel comes in and says that this picture of "spacetime" is wrong. And it shows the correct picture to be this :

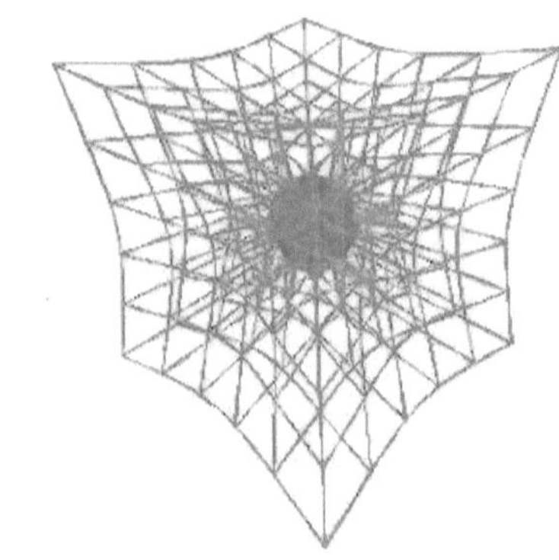

Figure (b)

The reel states that it is because space is 3 dimensional. And on the headlines of the reel, they called it "spacetime". And still more wrong is that it states that the image represents curvature in 3-dimensional space. Look at image (a), and understand that when you get the curvature in a 2-dimensional membrane, it gives you a 3-dimensional figure. And so the entire figure (a) is a 3-dimensional model created when mass curved the 2-dimensional spacetime. If you have a one-dimensional string and you curve it then you get a 2-dimensional loop. When you curve a 2-dimensional membrane, you get a 3-dimensional manifold. And when you curve a 3-dimensional manifold, you get a 4-dimensional hyperspace. So if a figure is a curvature in 3-dimensional space, then it is impossible for us to even think about it. Since our mind can think of more than 3-dimensions.

So a curvature always takes you a dimension higher. Figure (b) is not a 3-dimensional curvature at all. And so it's wrong to call it 3-dimensional spatial curvature.

So, since the spacetime is 4-dimensional, a curvature in it always makes it go a dimension higher. So spacetime curvature is actually a 5-dimensional geometry and the image (a) is a 3-dimensional analog of it. The image (b) does not represent the curvature at all. And so forget about it. Every physics textbook mentions the spacetime curvature in Figure (a) as the correct one.

So just to prove the concept wrong, I had to justify so much. And the reel I saw had around 10 Million views and around 130k likes. And a few months after I found no trace of that reel or reels like that. And once again yet I didn't see any attempt to correct the misinformation.

This misinformation misinformed so many people and diverted them from the true information.

So the conclusion is that the fame (like or views) of a reel doesn't make it conceptually right. It's just a

90-second video, where people are drawn to misinformation and people gulp it without thinking any further about whether that really makes sense. And if the reel favors the audience's favorite topic, the audience gets fooled still further. And that's what I said at the beginning of this book, "That, misinformation has its own Ph.D.". It's like, a baseless building. This is a waste of time. Stay alert and never get fooled by such influencers. The best way you can do that by is listening to a physicist + philosopher. Believe me, he knows everything. Many great physicists in the world turned towards philosophy when their questions about the universe were not answered. And so they had to become a physicist + philosopher to understand the universe. Those people who think that the entire universe including its origin and everything can be understood completely by physics alone are fooled.

Understand, that physics is a beautiful subject. And maybe it can explain 99% percent of the universe, but never full. And it's inevitable to turn to philosophy. A philosophy that has no contradictions with the physics we study, in fact, which supports the physics and physics supporting it. I have seen philosophers who claim any crap to explain their philosophy. Their philosophy contradicts the solid evidence. But the philosophy of a physicist + philosopher is always supported by physics.

The answer to the question of whether we are alone in this universe is a physics + philosophy conclusion. And of course, the philosophy is backed with physics. And physics which I am going to talk about is very fundamental and has no exceptions in this universe.

EIGHT

Science Fiction Movies and The Freedom Of Information.

The science fiction authors take a huge advantage of the freedom of information. The information has nothing to do with fact or fake. For example, a sentence "Not every action has an equal and opposite reaction". Now despite the statement being so wrong, it has a form of information. So information has nothing to do with fact or fake. It just can be anything. It can be even like a human breathing in water. It can be the violation of the laws of thermodynamics, it can just be anything. And when a particular piece of information becomes everyone's favorite, it becomes a motivation for fiction writers to produce content on it. Now what makes people crazier is the support of science. And if science says that the fictional information they like is theoretically possible, fiction becomes science fiction. It can be a film or anything. And the entire generation gets inspired by that. This also inspires the young generation for science and technology. Movies like Interstellar even inspired me as a kid.

But this is good only till the particular level of physics. Or till the topics that are only physics-based. As there are some topics, that are physics + philosophy-based. And when a sci-fi writer ignores the philosophy part, of such topics, he is also ignoring the physics. The fact that a particular topic is physics + philosophy based is because philosophy and physics have a meeting ends in it. It's a point where both physics and philosophy are in favor of each other and support each other. And by being ignorant of that point one is ignorant not only to philosophy but also the physics and then one comes to some bizarre conclusion. In answering the question "Why we are alone in this universe" a main physics point is ignored along with the physics-supported philosophy.

And so it might seem that the fake sci-fi has some physics and logic in it, and possibly it even be true in the real universe. But that's just fooling you since it ignores the main points from both (Physics and Philosophy). Such information can fool everyone. It's like faking pizza by its size by cutting out the straight slice in between and then the customers get fooled. When science fiction says that we are not alone in this universe, it is cutting out the physics and philosophy supported by physics that contradicts this. And thus, so many people are getting fooled by this.

NINE

The Bizarre Way of Making Conclusions.

If I give you 10 pieces of information about a particular topic and tell you to conclude the conclusion with one of 10 pieces of information unread, then would that be the relevant conclusion? The answer is, it can be right and it can be wrong, based on the unread piece of information. So final answer needs that one piece of information to be read. I choose such conclusions that are supported by all pieces of information available and do not miss even one.

Well, however obvious it might sound, most people don't do that. People forget, ignore, or hide a piece or pieces of information to prove what they want you to believe. Such ignorance can lead one to bizarre conclusions. And by giving such conclusions a limited touch of fine physics, anyone get fooled.

I see so many reels on social media platforms saying in confidence that we definitely are not alone in this universe". And now the misinformation is told as fact. This is assumed so much as a fact that fooled communicators now are finding complex alien structures in space.

I watched a reel on social media that was talking about some mysterious star whose brightness was fluctuating on a regular basis due to some parameter not yet known to scientists. I was surprised by one of the conclusions of the video saying that the star could be surrounded by a revolving Dyson sphere. By doing this, the communicator is taking misinformation to a Ph.D. level. And so he is fooling fooled himself. This is just a waste of time and precious life force. I am yet to explain to you why we are alone in this universe.

I have convinced many people about this. Many of them argue against this by forgetting and ignoring the pieces of information. And they jump from one piece of information to another. At one point, I was tired of doing this, And at another, I saw some other space communicator fooling the people again and again with the same thing. And then I decided that this topic now needs something greater than a reel, comments, tweets, and an Instagram or Facebook post, And so I written a book on it.

So now let's grasp some knowledge and move further in knowing why we are alone in this universe.

TEN

Why Are We So Special?

So much in the scientific media is humanity as observers are described as insignificant. These scientific media have a limited knowledge of physics. Since they don't know quantum physics. Having limited knowledge is the same as ignoring the facts. And when one ignores the facts, one can draw any conclusion. A stupid conclusion that spread among people with still limited knowledge. Throughout this chapter from Observers I mean Humans; The Intelligent Life.

Well, let's say we have a hypothetical coin called the Universe. On this coin heads mean "No

Existence of Observers" and tails mean "Existence of Observers". And imagine whenever this coin named universe is tossed, it always falls to the tails. And no matter how much you try, it always falls on tails. Even if you try a Googol 100^100 times, it always falls on tails. Is this still a matter of chance? For example, you see an apple falling under gravity and no matter how many times you try, it always falls on the ground. Then, will you still believe that it's just because of chance? No, we say that there is something that exists that makes the apple fall always downwards and so we call it gravity. Similarly, there must be something that is making our coin called the universe to choose tails no matter how many times one tries. It's nothing but the "observation".

People say this universe is just due to chance. They say that the universe chose the value of pi, the fine structure constant, speed of causality, and all other fundamental constants (in total 26) to a very very high precision just by chance. These type of people have limited knowledge of physics and make bizarre conclusions.

The light has two natures, one is waveform and the other is particle form. The light as an entity shows wave-particle duality. But it always chooses a single outcome when it interacts with something. During the process of the photo-electric effect light always shows particle nature. It does it every time. So, who tells light to always behave like a particle during the photoelectric effect and behave like a wave when it enters the atmosphere making our sky blue? If light behaves like a wave during the photoelectric effect, no plants and plankton would have survived, not even existed. And so no life. This is a practical example of our coin named universe. And there are an infinite number of examples like this. For those who say this universe is just created by chance, they must answer why our coin called universe always lands up at tails no matter how many times it falls.

If light is chosen to be a particle when it's in the atmosphere, the sky would not be blue and the oceans wouldn't reflect blue light. The amount of blue light reflected by the earth determines its temperature and is an essential factor in sustaining life.

Not only light but even matter are waves. So who tells matter to behave like a wave at a particular time and not at others?

There is an experiment in quantum physics called the Schrodinger Cat Experiment. In the experiment, the fate of a cat is explained to be the outcome of a particle choosing to behave like a wave or not.

So everything has a particle and wave nature. And the thing that chooses what everything should behave in the form of is nothing but an observation. In quantum quantum physics, it's a fact that observation collapses the wave function. And the one who makes the observation is an observer. This is called the observer effect.

There are philosophical questions like "Is there a moon when nobody looks?". This is an exaggerated example of the observer effect. In reality, observation is nothing but the manifestation that makes the universe support the existence of observers. And this observation is from nothing but the observers themselves.

People who believe that a universe just exists by chance don't know but the whole probability theory is an outcome of tails chosen by our coin named "universe". So if we have heads and tails and if tails always fall no matter how many times one tries till eternity, then the heads is just irrelevant information of the coin. In fact, the heads should not even come in the sample space if there is a solid 100% percent chance of tails landing up. So in such a case heads as an outcome is just non-existent. The probability theory is under the control of the universe itself. Think deeply about this. This may not be sensible at first, but deep thinking will make sense. And so the formation of the universe in this way is not the game of probability at all. This universe is finely tuned by observation in the sense that the observer and the universe both co-exist.

"No phenomenon is a real phenomenon until it is an observed phenomenon."

-John Wheeler

And since from observers, I meant intelligent life. Our original question of why we are alone in this universe translates to "Why are we the only observers in the universe?". Now let's turn to the final chapter of this book. And get our question answered.

ELEVEN

Why Are There No "Alien Observers" In This Universe?

I once read an article wherein it was said that the James Webb Telescope found city lights on some planet. The article annoyed me because as I explained, it was ignoring the known facts. And by ignoring the facts, one can come to any crap conclusion. Even a single piece of fact can prove a conclusion built on a thousand facts wrong.

There are billions of solar system-like stars system in our galaxy which have a habitable planet and we have many planets that are better at sustaining life than Earth. And there are trillions of galaxies in our observable universe. So that means intelligent aliens must exist, right? Crazy Science fiction fans in fact must be waiting to be killed by intelligent alien species assuming this. But sad for them and the other scientific media who love intelligent aliens, this universe does not work that way. It's now the time to get out of delusion. As always, these people lack knowledge about our universe.

But one must understand that the creation of intelligent life is just not a game of having habitable planets in billions of star systems in our galaxy and billions of such galaxies in our universe. It is also the game of our savior like Jupiter, it's a game of Saturn, it's a game of the moon and its proximity to us, it's a game of every planet and theoretically every star. Not Astrologically but Astrophysically speaking. If you are not convinced by what I just discussed, then let's dive further in this.

There is an argument in physics and philosophy that asks "If the universe is created by observers, then why was so much of the universe created if it could have created observers directly". This argument is called the Boltzmann brain argument. This argument asks why is this universe so symmetric but still anti-symmetry.

The Following highlighted sentences are the main part of this book, Read them carefully:

"Our universe tends towards symmetry". It does everything in a way that least non-symmetry is carried out. This is also a well-known fact of physics. This is the fundamental property of our universe. When the universe has to create some complex result, it will do so while maintaining the highest symmetry and creating lesser non-

symmetry. The formation of life is a very very complex process to carry out. And the universe does it creating the least non-symmetry. The creation of intelligent life is the highest non-symmetry ever created by this universe. And when it does it, it makes sure that everything goes as symmetrical as possible. This universe always chooses and "tends" towards symmetry. The creation of the universe is one of the goals of the universe and the observers have manifested the universe. So, the universe achieves that goal by tending towards symmetry. And once the observers are created, it still tends towards the symmetry. Creating more than one habitat for observers is creating so much non-symmetry than ever. By doing this it is that the universe is tending towards non-symmetry. This is not possible in this universe. The universe tends towards the least symmetry for the creation of observers just because it's manifested that way by the observers. And once the observers are born, and the goal of the universe is achieved, it creates no other less symmetry due to its property to always tend towards symmetry. There are trillions of galaxies in the observable universe, and each and every one of them has billions of star systems that have planets like Earth. This creation is nothing but the universe's way to carry out (symmetric way) the very high non-symmetry of the creation of observers by simultaneously making the process more and more symmetrical. The creation of another habitat for observers means the universe tending towards non-symmetry which the universe and physics do not go for after already creating the necessary non-symmetry with which the universe co-exists. So the universe only creates less symmetry when it's the goal of the universe to do that. And what creates the goals of the universe is the observation by the observers. One of these goals is the creation of the observer itself. This universe works this way. In this way, the universe has carried out its goal manifested by the observer while making sure that everything goes with symmetry. And since

the universe always chooses symmetry, it doesn't go for creating another habitat for the observer in the same universe. And that's why we are the only observers in this entire universe. However, the alien observers can exist in a Parallel Universe.

Similar to the Boltzmann Brain argument, the universe could have made only the solar system and so observers directly than creating billions of star systems in a galaxy and trillions of such galaxies in the universe. Then Why it created so much just to make a solar system, which we are a part of ? As I explained this is because the universe always carries out everything with the highest symmetry. And so if it wants to create an earth-like planet and solar system, it would do it by creating a thousand trillions of them. If it wants to create a sun-like planet, it will create trillions of them just to form the sun. The process of creating more symmetry is like creating a magical result with a process that seems less magical. The symmetry makes sure that the laws of physics everywhere are the same, wherever you are in this universe. In fact the universe is so symmetrical that the formation of life (high non-symmetry) is carried out in a process that seems so less magical. In short, the universe creates magic in a way that the process of creation of magic seems unmagical. But the result is always magical.

The universe creates no extra non-symmetry again since it always goes for symmetry. The creation of another habitat for observers means very high non-symmetry which the observer-dependent universe has to carry out. The reason why Alien observers don't exist (As discussed).

Also note that in this book as one sees on the front page of the book, from aliens I meant intelligent aliens. At the same time, there is a very very minute chance of having alien life in very basic form like bacteria and other microbes in our universe. But not the alien observers.

I want to share why this universe always chooses symmetry over anti-symmetry when everything can be just created so easily and within a second when carried out with anti-symmetry. I asked this question to the expert in spirituality and then he replied, "Just so that everyone can surrender to god".

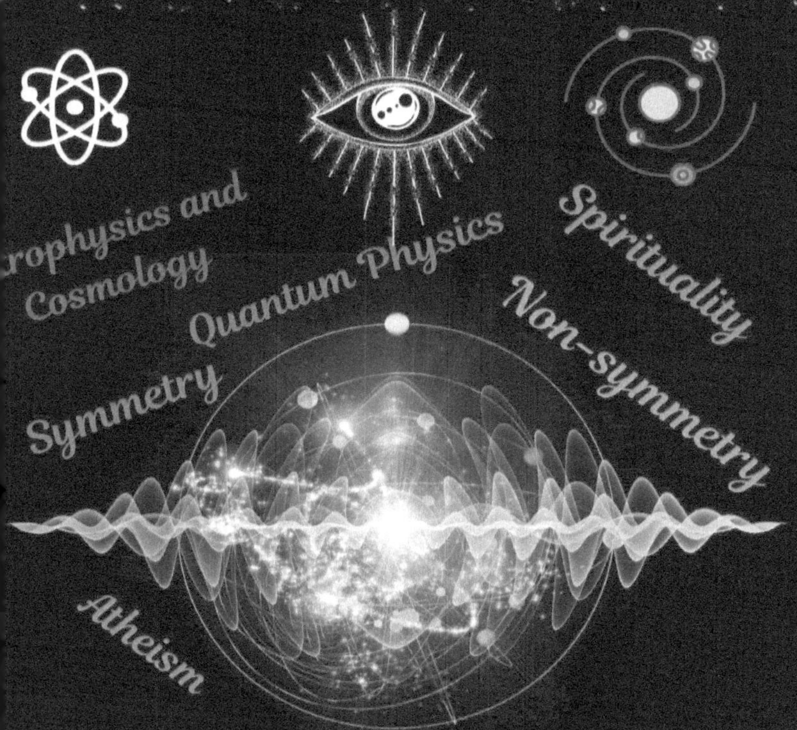

THE DIVINE DILEMMA
UNDERSTANDING THE TRUTH
FROM AN ASTROPHYSICIST, COSMOLOGIST, ATHEIST, SPIRITUALITY AND QUANTUM PHYSICIST.
Astrophysics and Cosmology
Quantum Physics
Spirituality
Symmetry
Non-symmetry
Atheism

THE DIVINE DILEMMA

TWELVE

The Information (2).

There are 8 billion people on earth. Each with their own expertise. Some people can explain god spiritually, but not according to science, while some can explain it by science but not by spirituality. Science does not talk about spirituality, the closest it can get to spirituality is its addressing of the observer effect of quantum physics. Almost every human is brainwashed to see the world in a way he has expertise in it. So, a brand or celebrity in just one of the fields of study alone can't comment on this topic and make predictions boldly. One must ensure that one does not believe everything a brand says. So, the topic of the divine needs the involvement of many other subjects as well.

Thus, to answer the question this book is based on, we need the teamwork of a person who understands spirituality, a Historian, an Atheist, a quantum physicist, and an Astrophysicist, and a cosmologist. And then by considering the conclusion of all of these experts, we can come on some conclusion.

In brief, this is how one is supposed to picture GOD.

Imagine a pack of iron fillings spread evenly on a sheet of paper. The paper let's assume is two-dimensional, on which the two dimensional beings live. These 2-dimensional humans have no sense of more than 2 dimensions. So there is only length and breadth on this dimensionality and no one understands the concept of height here as the 2-dimensional beings. Now let's say we pass a 3-dimensional magnetic field through this 2-dimensional world. Doing this, the iron fillings interact with the magnet and show some structure (of course we assume here that the example is conducted under the gravity of the earth for the sake of simplicity this example gives). The structure being formed is two-dimensional. For the 2 dimensional beings, there is no apparent source to know the deflection of these iron fillings on their 2-dimensional universe. This is due to the 3-dimensional nature of magnetic field which their mind is unable to think of. Also, their law of physics can't explain it. Because even their physics is based on the 2 dimensions. And so they can only get a limited idea of a 3-dimensional magnetic field based on the pattern the iron fillings make in 2 dimensions.

So, in early times God was known to be living in the sky. But realistically it means hyperspace. Hundreds of years ago people used to call a comet a demon of fire. From this, we understand that for the sake of logic and science, we should not take these words literally. And for this, we need a historian to understand the ancient scriptures describing GOD. We need a scientifically educated Atheist.

Spirituality or the science of spirit is extra-dimensional, and to comprehend it at a three-dimensional scale we chose a person who understands spirituality and can relate it to the real world. A quantum physicist helps us understand the quantum theory that joins the gap between spirituality and science. We need an Astrophysicist and cosmologist to help us to stick to our observations and evidence of science.

So the topic of "Does God Exist?" Should be a topic not just limited to an Astrophysicist and cosmologist.

So this book will contain the debate among all of these experts.

THIRTEEN

The Conversation

Throughout the conversation, the historian confirms that a significant part of history is written in the form of a story. In the form of a story because of our misunderstanding of the event or to communicate an event in an interesting way to the general audience. And these stories are not to be taken literally, and they have their own real-world interpretations.

Throughout the conversation with an Atheist, it is understood that atheists always seek some physical evidence. So they always expect magic to happen to consider it as evidence of God. They verify everything by science, which in fact is always preferred and a good sign of an educated person.

According to atheists the God is and was just an object of worship of help seekers and nothing more. They believe that people have a blind faith to give them a feeling that there is someone to help them.

Throughout the conversation with an expert in spirituality, it is understood that a soul is a fragmental spark of God. And the nature of the soul is extra-dimensional. According to the expert in spirituality, believing and understanding God are totally different things. One can believe in the existence of someone or something and still misunderstand or not understand it at all. According to them, there are only a limited number of people who understand God. The expert in spirituality doesn't deny science. They just believe that science or physics is not absolute truth, and it's just a small fraction of the energy of the god.. According to them, the reality most think is real is an illusion. According to them, the universe is created under the cause of God. According to them, there is a joining gap between science and spirituality just like the pattern created by iron fillings joining the gap between understanding the 3 dimensional magnetic field and their science of 2 dimensions. This is where one of the most discussed topics in quantum physics comes in.

Throughout the conversation with the quantum physicist, it's confirmed that reality at its basic level is nothing but a wave. A wave which has its peak and trough. But what keeps a quantum physicist in awe is the question of why these waves have a peak for a particular value and a trough for the other. And as we scale up, these peaks become sharper and troughs vanish. So looking from a basic scale, some cause is causing the wave to peak at a particular value and trough at another. So at the very basic level, the reality is nothing but a choice and before that choice is made, everything is like a sample space of possible choices that can be made. To understand why a particular choice is made, many Physics Nobel Prize winners and great physicists sought spirituality to understand this.

Throughout the conversation with astrophysicists and Cosmologists, it's understood that our universe is more than 13.8 billion years old. And humans have been on earth not more than 3,00,000 years ago, which is far from a billion. And still far from the 13.8 billion years age of our universe. According to an Astrophysicist, the planets and stars many light years away from us don't even care that we exist as humans here. And they say that the realization of God is by Humans which are very lately created after the birth of the universe. And so, how can we still say that God created this universe?

Note that this book is not a religious text or book of spirituality. The book just focuses on finding the truth with the help of science and understanding the opinions of various experts.

Above mentioned paragraphs were the summary of opinions we gained from the various experts. Now let's dive further into understanding the truth by converging the knowledge we summarized.

FOURTEEN

Understanding The Truth.

By reading the summary in the last chapter we are here to boil down the truth. First, we have to understand what it means by the creation "under the cause of". Let's take a fruit, the most important part of the fruit is the seed inside. The part outer to the seed nourishes the seed until it is ready to germinate in the soil. So, it's always the outer part of the seed of the fruit that comes first to nourish the seed even before the formation of the seed. So this is understood as " The fruit creates itself under the cause of the seed". Even If the outer part of the seed comes first, it's been created under the cause of a future seed. The same applies to the egg. When a bird delivers an egg, the life begins to form after a certain period of time in the egg. So it's egg that comes first and then the life inside it. Here it can be said that the egg created itself under the cause of the life inside it. So life inside the egg is the creator of the egg is a very direct way of saying that. Similarly, the universe is created under the cause of the fragmental spark of the God. That is a soul. And a human being is a soul. Here it's not said that a human being is a God. The deep knowledge in this subject can be achieved only in the subject of spirituality and the knowledge of the self (soul). This is not discussed in detail here. So, the universe is created under the cause of the fragmental spark of God, a soul. This is an important piece of knowledge for those who say that God makes no sense because humans were very recently born in the universe. The phrase "under the cause of" is important here. Also, just like the outer part of the seed is important for the creation of the seed inside, every star and planet is and has been proving as an important part of the creation of Humans who as a soul is the fragmental spark of the God. And just as a fruit remains nutritious even after the development of the seed, the stars and planets continue to remain in their process of stellar and planetary evolution. While the stars and planets near to earth still play a vital role as an outer part of the seed of the fruit.

In quantum physics, the one who collapses the wave function is called an observer. Same there, it doesn't mean that if the observer doesn't physically look at the moon, then the moon may or may not exist. The moon is still there even if the observer does not physically look at it. During the new moon day, no one can see the moon, but that doesn't mean the moon is in the tug of war of existence at that time. In the observer effect of quantum physics, they mean that the wave collapses in the cause of the observer. If a moon disappears, we can't expect life on Earth. Which is against the cause of the observer. The physics of quantum physics is only applicable at atomic and subatomic scales, the wave function of the moon is narrowly spiked. Ironically not even to consider it as a wave. So even the laws of quantum physics are designed under a cause of observers. So the observer effect doesn't mean that actually looking at a wave collapses it. It rather means that the wave collapses and chooses a result of collapse under the "cause of observers". So at the quantum level, the wave function chooses a particular value just under the cause of the observer.

Throughout the conversation with atheists, it's apparent that they are ignorant of the important branch of physics of Quantum Physics. They think that reality is not an illusion but solid physics. In this chapter, we refuted the claim of cosmologists and Astrophysicists that the conception of god came after humans arrived. This will be discussed in detail throughout this book further.

In the next chapter let us understand why there must always be the existence of a process of "action under the cause of". In other words, why does a mature seed not create itself from the quantum field directly, and why must there be some creation following it, under the cause of it? Or why can't we create humans or observers (the final cause) directly from the quantum field rather than creating the billions of galaxies, trillions of stars, and the entire universe for it? This argument is called the Boltzmann Brain Argument. Let's now in the next chapter see why the seed does not create itself directly from the quantum field, but rather creates a flower and then fruit and then manifest.

Symmetry And Non-Symmetry Of The Universe.

The atheists ask for the evidence of God. And when asked what evidence, they want God to show up himself in the form of a magician. Just like he must be able to lift things up without any medium and whatever. But let's now see what does magic mean.

magic

noun

the power of apparently <u>influencing</u> events by using mysterious or <u>supernatural</u> forces. "suddenly, as if by magic, the doors start to open"

So, the term magic means doing something supernatural according to the dictionary. Now let's understand what magic exactly is.

The term process is about taking an initial state to the final result. This can be done in two ways, symmetrically and unsymmetrically. When done unsymmetrically in a symmetry tending universe, the process is called magic. And the action is called Supernatural. I mean, maybe if a parallel universe tends towards such non-symmetry, doing things symmetrically would be magic (Ironically speaking).

Our universe always tends towards symmetry. And so every action that takes place here happens with the highest symmetry. So if we have to create a seed, then we have to create a universe, a solar system, the earth, the plant, the flowers, and then fruit to give birth to seed. This is the symmetric process. For an Atheist, evidence of God is his power to perform a process with non-symmetry.

And if God does it with symmetry, they call the process "Physics" and deny the carrier of the process.

So the question is why are atheists partial towards creation carried out with non-symmetry and not the symmetry? Maybe in the parallel universe where everything is carried out with non-symmetry, the atheists there demand to see god performing actions with high symmetry as the evidence of their existence.

Hiten Shelar @HitenShelar · 12 May
In the universe where everything can be created unsymmetrically the atheists are the ones who believe in God.

Read that again.

So if Atheists ask for evidence, they have to clarify why they choose creation with non-symmetry over the creation with high symmetry as a proof. This is the failure of the argument of Atheists.

So can God create non-symmetry in the universe where everything is carried with the highest symmetry? The answer is yes, but that would be equivalent to removing a glass from the base of the glass castle. And if did, the entire glass castle would shatter into pieces. So it's either you prove it to an atheist and destroy the entire universe or simply show the opulence through creations with the highest symmetry and carry out the purpose this universe is born for in the first place.

What Atheists want God to do as the evidence:

Initial State   ——————————————→

What God Does:

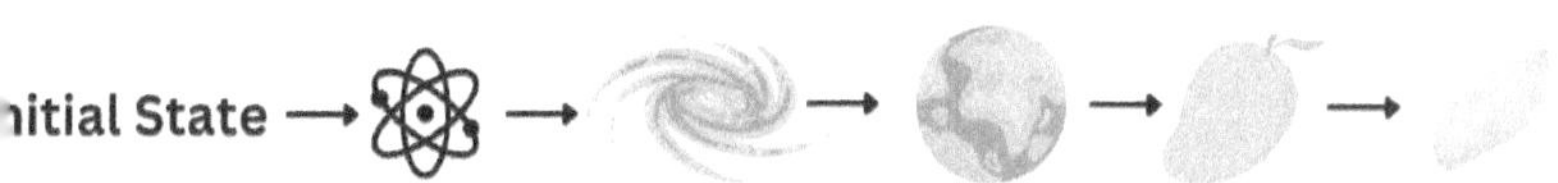

The above mentioned images depict the difference in creating with high symmetry and with non-symmetry. The first image shows the creation of a mango seed with non-symmetry and the second one with high symmetry. Both of these processes are equally appealing as the evidence of God. Atheists don't have a reason behind their partiality between symmetry and non-symmetry as the way of creation by which God creates in this universe. All one should focus on is the final outcome.

From this, we understand that Atheists have no rational arguments supporting their belief. The mango seed in this chapter is a metaphor used for the Observers.

SIXTEEN

Good Or Bad Nature Of God.

One of the other reasons atheists deny the existence of God is the apocalypse happening in the world. Since medieval times, people have misunderstood god as an object of help. According to the knowledge of self and spirituality, God is the object of devotional service. There is a difference between looking at God as a helper and as an object of devotional service. After receiving help a person will forget God, but once engaged in devotional service, the devotional service goes on for eternity. Eternity because the soul is eternal or immortal. A wise man said that God is like a gentleman, he needs to be invited, and the invite itself is a type of worship post in which the devotional service starts. When one asks for help from God, that is not an invitation. The real invite is followed by a devotional service.

That's why it's said that go to God first not at last. Since at last, when the battle is lost, the thing that is asked is help. But when one goes to God before his endeavor with full faith, that is an invite. And then the battle is fought in full devotional service.

If a Human is a material manifestation of a soul and a soul is a fragmental spark of a God, how is it still said that God created us? Since as a soul, a human is already a fragmental spark of a God. So Does that mean a Human is a God? Well, this is not said here. What is said is that humans are the manifestation of a fragmental spark of God as a soul.

There are three things existing, soul, mind, and body. A soul is the absolute truth. Every human has been given the independence to act in a way he wants as soul, mind, and body in the material spacetime universe that was created under the cause of his manifestation. The same material world also acts as a way to perform devotional service to God. Since every human has a little independence to act in the material world, he sometimes forgets that he is a soul, and sometimes he forgets that he is a mind and he just thinks that he is an Animal. Now when the word "Forget" is used, the question is who is thinking here at this point?

There are 3 levels of intelligence in a human, the intelligence of the body (brain), the intelligence of the mind (conscious + subconscious mind), and the infinite intelligence of a soul. And so if one forgets he is a soul, he acts under the influence of the mind and body. Mind is very wild in nature, at one moment it can be very happy, and at another, it can be very sad, it can self-betray and as well as it can help oneself. The mind can be the greatest enemy and the greatest friend. People who forget that they are the soul and mind, act under the influence of the body.

The influence of mind and body is good only when it acts to nourish and feed the soul. And when the soul gets the feeling of nourishment, that's a gentle or very basic taste of devotional service. And when the body, and mind together work in getting more of this taste to a soul, then as a soul (Absolute truth), body and mind together, a person is doing the devotional service unto God.

Now let's talk about Good or Bad. Let's say two humans are having a fight. Now, one might ask if a human is nothing but a soul and a soul is the fragmental spark of God, how can two sparks of the Supreme God fight and hate each other? In fact, this fight is what we have been doing as humans since the Stone Age for food. In fact, this is been happening throughout the evolution. So what is happening here? If God is good, why is this bad happening? Well, the act of fighting and hatred was just the nature of mind and body. In this chapter, it's been discussed here, how a human ignores his soul and acts like mind and body or just body. And depending on the mood of mind or wildness of mind and body, self (soul) hurting actions are conducted. If the mind and body are trained through devotional service, they never perform such actions. Humans have been fighting since the Stone Age for food. This shows the ignorance of the soul and mind back then. In fact, when the texts say that Adam and Eve were the first humans, they mean a human who knows that he or she is a soul. According to Charles Darwin's theory of evolution, back then there could be humans. But the first real humans Adam and Eve only counted when the humans first realized that they are the soul.

We talked about the symmetric and non-symmetric creation. The Big Bang, the forming of galaxies, dark matter, black holes, etc, and also the earth, evolution, and everything are just symmetric creations aiming at creating or manifesting a soul in the form of a human (Adam and Eve).

Talking about Good or Bad events in life or society there is no doubt that we as a soul, minds, and bodies are the only creators of our life. If one doubts this, one must educate oneself with Epigenetics, quantum physics, neuroscience, and spirituality about how we become a particular type of person attract things to us, and design our life.

A soul unlike the mind never gets fooled by anything since it's the fragmental spark of the god. The soul is infinitely intelligent. So if a person ignores that he is a soul and does self (soul) hurting and his mind and body don't lament over it, his soul still knows everything. So the word "ignorance" would be more accurate than "forgetting". So when a soul is ignored, and a person under the influence of mind and body performs such sins with no lamentation, the soul suffers. One can't at all ignore that his soul is suffering from this, the effects are always manifested for them. As a soul, no human can ignore this suffering of the soul. This is their world. It's called Hell. A person who always engages himself in actions that make a soul happy but at a limited level creates an environment for his mind and body called heaven. A wise man said that heaven and hell exist here itself. The statement is very true. Some texts say that heaven and hell are after death. After death, the human is present in the form of a soul. A soul is immortal. And if a soul goes to hell or heaven, that means a fragmental spark of God goes to hell or heaven. How can a fragment of God go in hell? This is rubbish. The heaven and hell are here on earth at the same place. The bad doers may look like they don't care about doing bad, but the intelligent always know that they are in Hell. So without concern about any person, first know by his thoughts whether he is in heaven or hell. One must always engage oneself with people who have engaged themselves in transcendental devotional service to the god.

Throughout this book, we have given only a limited amount of knowledge of self. Since this book is for a different purpose. This chapter was just to prove one of the Atheist's arguments wrong.

SEVENTEEN

Atheists And Flat Earthers.

Three kinds of Atheists exist. One who doesn't know what is God, one who knows but denies the existence of God due to his limited knowledge, and the other one who knows that God exists later but can't accept it because of his pride for his Atheist group.

Similarly, in flat earth society, there are three kinds of flat earthers, one who thinks the earth is flat and doesn't know its sphere, one who denies spherical earth because of his limited knowledge, and another who knows that the earth is flat but still can't digest it in the sake of their Flat earth society. Here is an amazing conversation between an

Atheist and God and Flat earthers and scientifically educated people.

An Atheist meets god in his personal form :

Atheist: I don't believe you are god, and if you are then show me creating a mango seed from nothing.

God: Here we go:

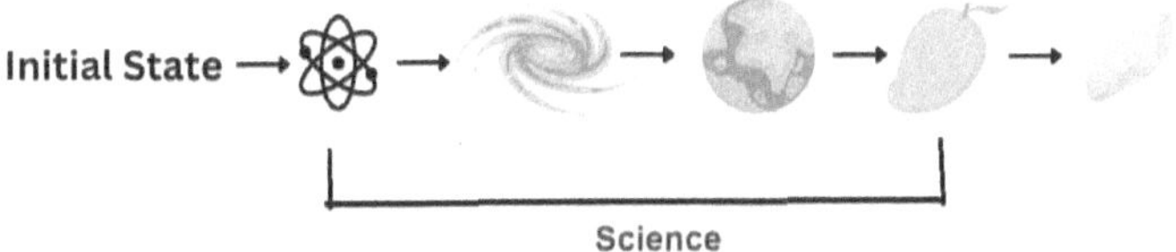

Atheist: Ok, can you create it without science?

God: Why is it that if I create it "without science" then only you believe me? In a parallel universe, the case is totally opposite, atheists there want me to create it with science and then they will believe me. Neither I will do there what I do here and neither what I do there will I do here. I can do it , but I don't prefer it as I will end this universe by doing so. And I can't do that just to prove you. I just created the mango seed with high symmetry. This is the proof. In this universe, everything is created with high symmetry.

Atheist: Ignores

<u>End of the conversation</u>

Next shows the talk between a scientifically educated person and a flat earther.

Flat earther: I bet Earth is flat and not spherical.

Science: I know you call the photos of NASA fake. Now hear this, according to physics anything as massive as Earth can't sustain the disc shape, it has to become a sphere at the end. This is how physics works.

Flat earther: You fake physics theories.

Science: We would get nothing by telling people that the earth is a sphere and not flat. Now the only way to end your flat earth society is to take every one of the flat earther to space and personally look at earth. Which of course is economically possible for NASA and so the government.

Flat earther: Ignores.

<u>End of the conversation</u>

So it's not that we are unable to prove an atheist and a flat earther wrong, it's rather that they are not "ready" to accept it. Note that this conversation only refers to a specific group of atheists. In fact, atheism in some quantity is required in today's world. This has been discussed in the next chapter.

EIGHTEEN

Atheism Is Not Always Bad.

Atheism is not bad until it one starts ignoring the truth. Many scientists begin their journey to understand our universe as atheists and they reach this point of knowledge. Where they start to understand that behind every why of science, there is a presumption "This is how it works". To understand this further, they study the very basic scale of the universe where reality is created out of choice from the quantum field, the quantum world. Then they study the choice maker, the observer. And then they study more about the observer, who is a human, a human that is nothing but a soul. And a soul is a fragmental spark of God. And then they start studying the ancient scriptures of God. In fact, this is a good approach. As there are various modifications made to the original knowledge of the self. These modifications many times have been seen spreading misunderstanding and self-hurting information. And it's common for people to use their logic and doubt it. Some wrong knowledge is wrong enough to be denied by everyone. Such people can't be called the atheists. Atheists just require the right knowledge.

Some people blindly follow the knowledge of self and the knowledge of God, without doubting it. This is very wrong. This puts one at risk of going in the wrong direction and creates hell for oneself. Thus the attitude of an atheist is needed. And after that one needs the right knowledge of self and the knowledge of God. There are many people who believed in God, then turned into atheists, and then again started believing in God. Again because they got the right knowledge. This type of atheism which takes one to a right knowledge of God is necessary. Doubting the scriptures is necessary, being an atheist until the right knowledge is necessary. Just that atheism should not be taken at a level that one ignores god even after receiving the right knowledge.

NINETEEN

Creator Or Created Under The Influence Of?

As we discussed early in the chapter there is a difference between doing something and something happening in some cause of. Doing in some cause of can be understood as the cause himself is doing, but he does not indulge himself in doing everything. Just like if the manager does work under the order of the owner, the work should be considered to be done by the owner himself. But at the same moment, the owner doesn't give a part of his time to do the work of the manager. Of course, not every work is done by the manager and sometimes the owner has to work as well, but only for limited work.

So if for example somebody gets fired from the job by the manager, the owner must be not considered 100% responsible for it. But since it has occurred under the cause of the owner. So making the owner responsible for every action happening in the company is partially right and partially wrong. Partially wrong because of the independence of how to perform the work of manager and every employee in the company. This universe is not made by the God, but instead being made under the cause of the God. But as a cause of the owner, the final result is the profit of a particular company. What Is the final result of the universe created under the cause of God are the "observers". So just like every credit is given owner to the company. The credit of creating the universe goes to God. And so its said that it's God who created this universe. This although an indirect truth, but still a truth and a fact.

TWENTY

What Happens When Non-Symmetry Is Conducted In The High-Symmetry Universe?

One might ask now, that if a god is all-powerful, why is he then not seen conducting non-symmetric actions? In our universe already, non-symmetry is getting created at a limited level. And what is asked here is the high level non-symmetry. The reason God is not seen   high non-symmetry being the nature of this universe. One can imagine this universe as the glass castle. Even if one glass is removed from such a castle, the entire castle shatters into pieces. Similarly, our universe is maintained at a specific limit of symmetry and non-symmetry. If non-symmetry is performed beyond this limit, the entire universe ceases to exist.

One can now question why only this amount of symmetry and antisymmetry is permissible in this universe. The answer is because of the purpose of this universe. So this universe is highly symmetric just so that the observers or human performs the devotional service in it. This a very straightforward answer. But as we go ahead in this book, one will see it more as a fact.

TWENTY-ONE

So Does God Exist?

(The Conclusion)

The above conversation doesn't target to prove who is right and who is wrong among the debaters. The debate is rather an approach to reach the truth by hearing from all.

By the above conversation, it's apparent that God exists, but not the way common people understand it. The science of God is discussed only to a limited extent in this book. There are things in ancient scriptures that obviously sound wrong if taken literally, but their intended meaning is not wrong. One must be patient and calm in understanding the intended meaning behind it. This was the natural way of communication back then. Maybe the description we write now of the present, which will be part of history thousands of years from now sounds wrong due to the change in the way we communicate today. Which of course doesn't mean that we are lying or it's just a story. Similarly, the terms worship, devotional service, and many other important terms are largely misunderstood by the majority

The Devotional service is always nourishing, there is nothing sad in it. It's the real world. The illusionary real world contains lot of sad things. In the real world, there is no sad reality. Now, the matter of divine worship is not described in very detail in this book. This book aims at a different topic. Note that, the majority of people largely misunderstand the devotional service. So before for knowing the true meaning of devotional service, a little bit of atheism is required in today's world.

Throughout the conversation in this book, we understand that the universe has been created under the cause of observers who are no one but the soul. A soul is nothing but the fragmental spark of the supreme. The main goal of a soul, mind, and body together is to do the devotional service to God. This is the prime purpose of this universe.

In that sense, we conclude that God does exist. He is not just the object of belief, he is the object of understanding, he is the object of worship

ABOVE THE
COSMIC
PERSPECTIVE

ABOVE THE COSMIC PERSPECTIVE.

TWENTY-TWO

The Cosmic Perspective.

The sentence cosmic perspective over the civilization refers to the perspective where humanity is seen from the perspective where we are not different from each other as the habitant of a pale blue dot (The Earth from the cosmic scales).

Our sun is a third-generation star born 4.6 billion years ago in the universe with an age of more than 13.8 billion years. In that sense, the common understanding would be that humans are just stardust. We, humans, are created from the same mass-energy that was born at the dawn of our universe. And our universe is infinitely huge it's expanding faster than light. And we have seen no signs of life at the very large scale of the infinite universe. We could even be alone in this universe.

When seen from this perspective humanity comes closer together. And the subject of discrimination comes to an end and everyone starts caring about each other.

The idea of cosmic perspective thus brings unity to civilization. Under this perspective the strange rules of our government, the rules of religion, our anxiety, depression, etc seem useless and of no worth and matter. So keeping this thing in mind we have to live for good and make this world a better place as intelligent beings. But there is something better than the cosmic perspective which is not just an idea, but the fact, the truth. Where everyone who doesn't believe in that truth suffers and those who act in that manner enjoy eternal happiness. This truth again, is what everyone likes to live in, but they are far from it due to their own independence of actions.

The cosmic perspective also has its bad sides. For example one can commit a crime commit a crime under the motivation that no star in the universe gives a damn about it. In most of the books on cosmic perspective, this downside is ignored. In the real world, a soul suffers when the mind and body misbehave or conduct any action that hurts the self (soul). This suffering of the soul can never be ignored. Many books on cosmic perspective ignore our role as the observers in this universe as highlighted in quantum physics. These key topics are largely ignored in those books. Considering our role as observers of this universe is a vital part of the discussion. This part of this book, takes our roles as observers into the discussion. And the reader now must watch how the cosmic perspective changes. To add, this updated cosmic perspective has no downsides or sad realities whatsoever.

Before we move ahead in this chapter, let's take a deep dive into the concept of hyperspace. The concept of hyperspace is needed to understand the knowledge of the self.

The Hyperspace

The concept of hyperspace is very very important in understanding the knowledge of self. Let's first understand the concept of hyperspace. The majority of people think that we live in 3 dimensions. Because everything around us has a length, breadth, and height. And the majority of people only believe in what their eyes see and minds can perceive. They think that this universe is 3-dimensional. But the deep study of the universe suggests us that the time is also a dimension. And the scientifically educated people know that this universe is 4 dimensional. Lets now get a picture of higher dimensions.

Lets take a point. A point that has no dimension at all. The zero dimension. Its not possible for mind to think about zero dimensions. So just for the sake of ease of understanding, just imagine it as a dot.

When two of these dots are connected we create a one dimension. Still not possible for mind to perceive, so the very rough picture would be to imagine it as a straight line.

Now comes the 2 dimensions. When two straight lines are connected, we can get a square, which is 2 dimensional. The two-dimensional object only has length and breadth, and there is no concept of height. Anything on your mobile screen gives a good picture of a 2-dimensional entity. Although even if it is just a light.

Now, the three dimensions. We know this dimension very well. Our mind can perceive this dimension. The three dimensions get created when we create two two-dimensional planes.

Now comes the 4th dimension, the unperceivable for the mind. When two cubes are connected in higher dimensions we get a four-dimensional cube.

When two 4-dimensional cubes are connected in higher dimensions, we get 5-dimensional cube. And so on.

The concept of higher dimensions (more than 3 dimensions) and higher dimensions itself is called the Hyperspace.

Most people believe only in what their body can mind can perceive. As the common man is ignorant of the spectrum other that visible spectrum of light. The mind and body is many times ignorant the real nature of the soul. The soul is a higher-dimensional entity. A human being is a soul, body, and mind together. So as a human being we just not limit ourselves to 3 or 4 dimensions. Since a human being is a soul, the nature of human is higher dimensional.

Let's have a two-dimensional universe and in there two-dimensional living beings. Let's say a 4-

dimensional living being poked her limb in such a universe. In such a universe, the finger would appear in form of a circle and not like a 3-dimensional limb anymore. This is because the mind of the people there can't perceive more than 2 dimensions. So the action of poking of the limb will appear as a circle oscillating its shape and size, based on the dimension of the limb. So just because the people couldn't see more than 2 dimensions, that doesn't change the real dimension of the limb in reality. The same way, the body and mind is just a small component of a higher-dimensional soul. Everyone needs to understand this. The soul is not inside the body. The body and mind are just the 4-dimensional small component of a higher-dimensional soul.

The picture of a higher dimension can be given through this story:

On a normal summer day Joe comes back to US after a long business work in Japan. This return time he decides to surprize his parents with a new bungalow in their home country US. The moment he was to surprize his parents with a new bungalow. The earthquake happens, The bungalow was cubical in shape with cubical shape rooms. In the earthquake, the cubical rooms begin to topple. And what lefts behind is the single cubical room. The other cubes just disappeared. When Joe cautiously enters the house to know what's happening in bungalow which just appears as a cube now, he gets surprized to see missing rooms in the undamaged state as he enters the house. But the house just appears as a single cube as seen from outside. How is this possible?

As Joe decided to check further, he climbs the stairs and finds the master bedroom above the entry way. Instead of finding the second floor, however he finds himself back on the ground floor. Thinking that the house is haunted, joe ran towards the main door. And instead of leading to the outside. The main door leads to another room. Joe gets faint. As he explore the house, he finds that each room is connected to an impossible series of other rooms. In the original bungalow, each room had a window to view the outside. And now all windows face other rooms and there is no outside.

Scared of what he witnessed, Joe slowly try all the doors of the house only to wind up in other rooms. Finally he studies, and decides to open the five venetian blinds and look outside. When he opens the first venetian blind, he finds himself peering at the Petronas twin tower. When opened the second venetian blind he was staring at the vast ocean. But upside down. Joe then opens the third venetian window he finds himself in Paris. Then he opened the fourth and he stares at the forest. Forest not on earth but on the earth like planet. After opening the fifth venetian blind, he was looking at the dessert from Mars. After haunting experience of the rooms of the house, Joe finally figures it all out. The earthquake had collapsed the joints of various cubical rooms and folded the bungalow in the four dimension.

From outside his bungalow looked as a single cube. The bungalow didn't collapse due to earthquake because, the joint between the cubes were rigid. And when the earthquake appeared, the rooms of *the bungalow folded into a hypercube (Higher dimensional Shape). Joe got no way to figure out how to get out of the house. Until a similar earthquakes hits back and the house return back the normal with one room damaged. Joe again tries to open the main door of the house. And now he finds himself at the backyard of same house instead of front. Joe was happy, at least he could get out of the house.*

Throughout the story we learn, that joe was totally unknown what was happening with him and his house. This what happens when a mind finds itself in the extra dimension. The human mind is not even able to think about the extra dimensions.

Hiten Shelar
@HitenShelar

The concept of Hyperspace is very much needed to understand the knowledge of "self".

8:12 · 22 Jul 24

The soul as fragmental spark of the God. It's extra dimensional in nature. And how can a person as a mind and body take responsibility to perform the action at that level. The intelligence of mind and body can't even think about extra dimensions. The only way the person as mind and body has to perform actions in full devotional service. These actions are the higher dimensional equivalent actions as they nourish the soul. And a person as soul, body and mind perform the devotional service.

TWENTY-FOUR

The Idea.

We everyone are in the search of truth. We constantly gather the ongoing information In our society. We debate it with experts and non-experts and we land at the conclusion. And thus every subject has its own conclusion. And it's commonly seen and observed that people change their opinion subject to subject. Most of people forget to involve one conclusion while building another. For every topic they have some conclusion. They can be seen in believing two opposite ideas on the same time, based on the subject of discussion.

So they have a separate conclusion for each and every subject, and they don't care whether these conclusions meet each other at some common point supporting and explaining each other. This is the ignorance. While believing in any conclusion, one must make sure that it supports and is supported and explained by every other conclusion. If you don't take care of this, you get fooled.

In inception, it is said, that an idea is like a parasite. If that idea is in the form of a question that makes one uncomfortable, the uncomforted doesn't go until the question is answered. This book is answering these questions through multiple topics here in this book. One of which is the notion of the cosmic perspective. The vastly known and understood cosmic perspective ignores many such vital points. This lands the believer of such a theory to false reality Here we are knowing the truth. Neither we have denied science nor any evidence. Since we know that the facts are ideas, it's not possible to ignore them to forcibly reach our conclusion.

TWENTY-FIVE

Do We Really Don't Matter In This Universe?

People with high self-esteem will say we do matter in this universe, while people with no hope in their lives will say we don't. But who is right in this case? One of the meanings taken from the cosmic perspective is that we are the inhabitants of a pale blue dot. And just because we are small in this universe, we don't matter. I mean, why is it that the small things don't matter? The life in the egg starts from the microscopic embryo. The entire egg is made just under the cause of that embryo. And there is a huge size difference between an embryo in the egg and the egg itself. It has nothing to do with the size. And so if a big thing is formed under the cause of the small thing, however small it is, the smaller thing shouldn't be considered of no worth.

Secondly, there is an important point missed by the conclusion of the cosmic perspective. Well take a minute or two and think what am I going to mention here. It is nothing but a fact that we are the observers of the universe. The observer of this universe is a human being, a human being is a soul. In the chapter on Hyperspace, the nature of the soul was discussed. This is a necessary point to consider. And all our conclusions need to take this into account. Ignoring this can land one at the false perception of reality.

So, while making any conclusions, one is suggested to make sure that the conclusion is not just limited to that topic. But also, it should provide a basis and support every other conclusion.

TWENTY-SIX

How Do We Matter In This Universe?

On a bad day, do the stars in the sky really care that I have a bad day? Is for any star, am I as great as I think I am? These are the questions that comes in mind when one thinks in the cosmic perspective. Where our mere assumption and actions appears to make no difference in reality.. And it's always society that brainwashes you to believe that your are dumb and useless just because you score less in the Sunday test.

And our actions just seem to be doing the material combination of the things here on this planet. And that doesn't seem to have a connection with a far distant star. So in such a frame of reference should a person think that his actions matter? How this all works? Let's understand this.

We are the observers of this universe. An observer is no one but a human being, a human being is no one but a soul and a soul is no one but the fragmentation of the God. A fragmentation of God that is higher dimensional. God exists in still higher dimensions. When the observer performs a work in the devotional service by mind, body, and soul, the process is called an "Action". Anything not done in the devotional service is not an action. Although it might seem like an action, but it's the illusion. When we act in devotional service, it's not just our body and mind (small component of higher dimensional soul) that acts, but the human as a soul, mind, and body perform this action. This is the true action. Now how am I going to trust this? One needs to study epigenetics, neuroscience and quantum physics to know how the act in devotional service (action) changes things around us. Or how when a human acts as not just a delusional mind and body but as a soul, body, and mind together. This is what happens when one acts in the devotional service. And for those who don't perform the actions (work in the devotional service), the change does not take place.

So we do matter as the observers of this universe. But our actions only matter when they are performed in the devotional service. Otherwise, those actions are just nothing but some material combination, of which the universe doesn't care about. The observer thus always has the freedom to behave like an observer or just a material combination of elements in this universe.

"The observer's actions are the actions done in devotional service, otherwise there is no action."

Two people perform actions, one in devotional service and the other without devotional service, both of them get's fruit of their actions. But one action is a real action and the other is simply the illusion. The fruit obtained by one who performs the action in devotional service is the only real fruit. Because the proof of relation between that fruit and the receiver of that fruit doesn't only limit itself to the material combination, but also from the tiniest atoms to the grandest scales of the universe. And so the person's action and fruit of action marks the real proof of reality. Fruit obtained with illusionary action is just a material combination of the atoms and nothing else.

When an observer under whose cause this universe is made performs the real action (work in the devotional service), the action and the fruit of action ripples off from the tiniest to the grandest scales of the universe. This is where the false conclusions taken from the cosmic perspective come to an end.

So if a human being or an observer performs the action is devotional service, does that show a cosmic effect like a supernova, formation of a new star, etc.? The answer is, that it shows much more than that. Hear this out, one needs patience to digest this. An egg is formed under the cause of life inside it, and the prime duty of life is to grow inside it. The egg has sufficient and right amount of material to give life enough nutrients it requires throughout its growth inside the egg. So the egg under the cause of life inside it, is always prepared for the life inside it even before the egg is a long time away from hatching. So, the universe is made under the cause of observers inside it. Where the duty of observers is not just to take birth, but to do the devotional service. With the help of classical physics and general relativity, we can accurately predict the evolution of a structure of some part of the universe. We know how and when the sun and other stars are going to die. And the rhythm of the star formation and its death, we study the neutron stars sending gravitational waves, etc. We figure out and predict, that things in the universe have to behave in a particular way. And why that particular way is because the fact, that the universe is pre-prepared for the prime purpose it has been formed under the cause of. Some not fully scientifically educated people, answer this question by saying the universe is the way it is because of physics. But they don't understand that the physics itself is the pre-preparedness. We have discussed what physics is in the topic "The Divine Dilemma" of this book. This is one of the reasons why I arranged this topic just after the same topic of this book. So just like an egg is pre-prepared with a plan to allow the life inside it to perform his action of growing, the universe is pre-prepared for the observers so they carry work under the devotional service aka "an action". This is what the universe is made under the cause of. And it is this pre-preparedness that we can't relate to our actions and that causes us to take out the wrong conclusions out of cosmic perspective. These wrong conclusions only apply to those who don't perform action. An action can even be breathing and eating. The action performed in devotional service is the real action and the fruit received is the real fruit. The signs of this action and the fruit of action are seen in a cosmic perspective in the form of the pre-preparedness of the universe. The pre-preparedness also serves as the support of the action and the proof of the result of the action. And for those who don't work in devotional service, the action and it's fruit is not counted from the tiniest scales to the grandest scale of the universe. And the fruit is just the material combination. It's just a waste of time as the observer.

Just like one cannot beat the river into submission and one needs to surrender to its current to use its power as one's own. The pre-preparedness of the universe is a flowing river, and when one performs the action in devotional service, he surrenders to the river and uses the pre-preparedness of the universe to achieve the fruits of his action. Such a fruit is a real fruit and the satisfaction from such a fruit is a real satisfaction. And as we discussed the pre-preparedness acts as support to the real action (action in devotional service) and as a proof of the fruit achieved from the action.

Just like a life inside the egg doesn't know why there is so much egg york and white of an egg in there. The science and mind hasn't still figure out, how the pre-preparedness of the universe works. So the life inside the egg can't call itself useless just because of the size of the egg, during life's initial stages, not knowing that the cause the egg has formed under is the life itself. Same applies to us, if we don't understand the pre-preparedness of the universe, just shut up. All this is to support our real actions (action in devotional service) and the cosmic proof of our real actions as observers of this universe.

This is the original. cosmic perspective.

TWENTY-SEVEN

The Interpretation.

On 5 June 2024, I posted the following message on X:

Hiten Shelar @HitenShelar · 05 Jun

People with no cosmic perspective think humanity matters in this universe.
People with cosmic perspective think humanity doesn't matter in this universe.
People above cosmic perspective think that humanity does matter in this universe.

This is true, but not literally.

Here the cosmic perspective means the widely understood false interpretation of the cosmic perspective and "above the cosmic perspective" means the real cosmic perspective which I explained in the last chapter.

Sometimes, the false cosmic perspective becomes better than no cosmic perspective at all. Just when a school calls a person a failure, the kid will always believe that he is a failure for his entire life, not knowing that the cosmos doesn't give a damn about the school's opinion. The interpretation of the false cosmic perspective teaches us how, toxic the world can be if taken seriously. The false cosmic perspective of course has it's good side I mean. When we look at the stars, the stars never care whether you are insulted by your boss and your so-called friends laugh at you. But not taking it seriously means also allowing everyone to repeat the same thing without any self-care. The true cosmic perspective teaches us to act in devotional service. So if the action to retaliate is the action in devotional service, the actions must be taken with no hesitation. Again, one cannot call every action as a devotional service. The action in devotional service is not explained here in detail, since this book aims for different things.

But still, to explain in brief, a human being as a soul, body, and mind is required to feed his own self (his soul) by performing actions that enrich the soul. The same soul which is the fragmental spark of God. This is the feeling of action done in devotional service. Allowing others to hurt the self (soul) is against the devotional service. A soul never hurts a soul, but the mind and body can. In such a situation, the retaliation becomes part of the devotional service. And at any cost, the action must be taken. And the effects of this devotional service as discussed earlier matters at the cosmic level in the form of pre-preparedness of the universe. Pre-preparedness for the observers of this universe (humans) to act in the devotional service.

So the true cosmic perspective teaches one to distinguish between what to take seriously and what to not. Acting in the devotional service is the only serious thing and rest is just the independence given to humans as body and mind.

The false cosmic perspective which is better than no cosmic perspective, misguides one to think that whatever he can do anything, the universe won't care. Reflecting on this, a person can commit a crime and say the universe doesn't care for this. But the true cosmic perspective doesn't allow this, as the soul can never hurt a soul. It's always the wildness of mind and body that sometimes can't distinguish between an action performed in and not in devotional service. Otherwise, a soul can never hurt a soul. If the mind and body does, it's the soul of a person that suffers, which can never be ignored by a human as an absolute soul. And it always suggests one to act in full devotional service. Acting in devotional service can never result in failure, since the universe is already pre-prepared for it. So thus, every action must be performed in the devotional service.

So throughout this topic of the book, the true cosmic perspective is explained.

TWENTY-EIGHT

The Role Of Technology.

This book talks about the truth. And it can not ignore the facts. Since we can't ignore what is really around us and is playing a significant role. And we need every conclusion to match with every other to come to the truth. Let's learn how the role of technology has been in the universe that is created under the cause of observers residing in it.

Throughout the history of man, the day since we lighted our first lamp. The technology has been a part of our day-to-day life. So hear this out. Technology has always taken mankind closer to his soul. The closer to the soul mind and body is, the more he knows God as mind and body, and so the more he tends to do the devotional service.

The technology has increased the number of ways a man can stimulate his mind to a higher level. Technology has brought down the wars. Today we have nuclear weapons, only two of which have been dropped on the civilized places. That's one of the saddest events in the world. Back in the age of kings and queens, somewhere in the world, a war used to take place every single day. In a single war, thousands and lacks of people used to get killed. Even today war happens but under a fear of the use of nuclear weapons from either side. This fear has reduced the frequency and extent of war taking place. So the number of people dying in war has largely decreased as compared to the past times (according to overall statistics). Again, the opinion is always against the use of nuclear weapons. Having a war, or starting a war in the first place is self (soul) hurting action, and that is not the action in devotional service. The war only becomes part of devotional service when one has to protect his/her soul from a wild mind and body trying to hurt his and another's soul. So from the very beginning, the war starts with a self-hurting action, which is never given any support in this book. This chapter on this topic just explains the role technology has been playing in the universe which is made under the cause of observers residing it. The observer is a human being and a human being is a soul. And a soul is a fragmental spark of the God.

Technology has increased the number of ways, to stimulate the mind. Under this stimulation, a man listens more to the higher authority of the mind, which is closer to his soul. The stimulants can be harmful. The harmful stimulants aren't recommended here. As said, everything must be done in devotional service. So there are good stimulants as well. For example, a mobile phone if used in a bad way can make a man a demon. While used in a better way one can connect a man to the world, he can find people who love him and his passion at the global level.

We as humanity are recommended to live together at the same time. This may not be possible if there are territorial boundaries on the earth. But the mobile phone has broken this barrier. Now one can be in long distance relationship with his or her soulmate. The freedom the mobile phone has offered is historical. So, technology has always helped man to get closer to his soul and has even added power to devotional service. So, there are numerous examples of how technology has taken mankind closer to their soul. So, the conclusion is that the role of technology in this universe is to bring mankind closer to his/her soul.

TWENTY-NINE

We Know Everything.

It would be an obvious conclusion that based on the size of our universe, we know nothing about it. The observable universe is 93.016 billion light years in diameter. And so we barely have explored our universe. So, with such a high confidence how can our conclusions meet the conclusions set by the unknown universe? To answer this, I would have to take back the reader to the concept of symmetry.

As discussed in earlier chapters, our universe is very symmetric in nature, and it creates everything carrying out a high symmetry. If the universe had created some non-symmetry then that part would be the part of the universe closer to the observers.

Except that the universe is very symmetric. So just like we can expect roughly the same number of gas molecules in any square centimeter or meter in ideal conditions, we expect the universe to be roughly all the same everywhere. Although except the habitat of observers under the cause of which, the entire universe has been created. So except for our vicinity, we can expect the same story of the stellar cycle and planet formation everywhere.

Every part of an egg is important in growing the chicken inside it. This is the preparedness of the egg, as it is made under the cause of successful growth of life inside it. In the same way, the universe is made under the cause of observers residing in it, and every part and aspect of the universe is pre-prepared to create observers and support their devotional service.

There are three levels of intelligence, bodily intelligence (brain), the intelligence of mind, and the infinite intelligence of a soul. Infinite intelligence knows the answer to every question of this universe. When we say we don't know anything about our universe, we actually mean that it is our mind that doesn't know. The intelligence of the mind is immense, but not infinite. We as a soul know everything. Now one will ask, why can't the soul reveal everything to mind? To answer this, one must understand that the soul is infinitely intelligent, and a super-intelligent person doesn't only know the facts but also knows how to and when to reveal those facts. One might say why will the soul not reveal that piece of information? The word information Is important in this question. Now one must understand that the only difference between the Stone Age and the present civilization is the piece of information. Back then we had everything, and it's just the piece of information that has brought such a tremendous change. The soul having infinite intelligence knows this and so also knows how to act and when to reveal what. But when there is a talk between two souls, the information can sometimes be communicated. However, for that a person needs to go enough close to his soul as a mind and body. So when one as a mind and body goes enough close to the soul, one knows this infinite intelligence of the self/ soul.

So we all know the answer to every question of our universe, but only as a soul right now. Again, being a physicist can be path of devotional service of somebody, so there is a good amount of support to the science community searching for the unknowns of the universe. Unknown to the intelligence of mind. As an infinitely intelligent soul, we do know everything.

THIRTY

Things That Terrify Humanity To Lose Their Faith.

There are some things or some events in this universe that might make us feel that we are not special. For example, the sun. The sun is expanding as it's depleting it's hydrogen fuel. And after around

7.59 billion years, the sun will engulf the earth. This is the point where one might think that the universe doesn't care about us. But this is wrong. As explained in previous chapters, the universe has been created under the cause of observers residing in it. And the entire universe the way it is, is the pre-preparation of the entire universe to support the existence of observers and their devotional service. So expanding sun is nothing but the pre-preparedness of the universe to support the devotional service of the observers. Again, the observer here means a human being. A human being who is a soul, and a soul that is a fragmental spark of God.

And it's not just the sun. There is a threat from huge meteorites as well. One thing should be again cleared that this is the pre-preparation of the universe. Again, to conclude every aspect and event of the universe is nothing but the pre-preparedness of the universe to support the existence of observers and their devotional service. So one must not be terrified to lose his faith in him and the devotional service. One must take care of this. In the real world, there are no "Sad realities" that do not exist. Word to word there Is no exception to this rule. This book based on its main topic of discussion, might not be enough and cover every doubt regarding this, but the conclusion stays rigid.

THIRTY-ONE

The Action And The Fruit Of An Action.

This is an important point to consider. Since the dawn of civilization humans have been engaged in action and the fruit one gets by it. Just like every action has an equal and opposite reaction in physics, every action is followed by the fruit of that action. This rule although seems to break in some events like the very competitive exams.

The best example would be the exam conducted by the most populated country in the world, India. The UPSC exam, which is a gateway for becoming IAS officer in India. Around 1 million students give this exam. The toughest part of this exam is that out of a million, only 180 (as per July 2024) become IAS officers. So the selection rate becomes 0.018 percent. Of course this a bad estimate, since we didn't consider the psychology of the students. So let's say out of 1 million students, 2,00,000 students are preparing seriously for the examination. And out of those 2,00,000 students, 40,000 students are super serious for the examination. And all of these 40,000 people study for 14-15 hours a day. Out of these 40,000 people only 15,000 with studies have a good knowledge about everything that is asked in interview. And so all of these 15,000 people are capable of becoming an IAS officer. But only 180 out of them make it. I mean still only 180 out of 15,000. Which just makes 1.2 percent. So one might say that 14,820 students out of 15,000 got no fruit for their actions.

So, what are we seeing here? And they were all the best of the best. Every action must be followed by the fruit of the action. The problem I raised here is not just common in the UPSC examination but in many other competitions as well. So what is going on here?

Once my friend asked me, "What could be the cruelest thing one can do to the world?". I replied," It's spreading the wrong knowledge of "self". The knowledge of self or soul which is the fragmental spark of God and the knowledge of the God is the highest form of knowledge ever.

The truth of this universe is that "every action is followed by the fruit of the action". And there is no doubt about it. But now how to convince our senses about this when the results show that fruit hasn't been achieved? At such a point man has three options. One is to pretend that he didn't work harder. Yes, I mean pretend. Second, to scare his soul by believing in the false knowledge of the self. The second option is the worst of all. This option puts one in hell. By accepting the second option one chooses hell. The third option is to not accept the defeat and provide one with the right knowledge of the self. By doing this one will get to know that his action has always been giving him a fruit unless and until he chooses the first two options mentioned above.

A gentleman once asked, "Who performs the action, the soul, mind, or body". I replied the soul is infinitely intelligent, he can create anything, anyway, anytime. So the soul is always convinced that it owns the fruit. But what one needs to convince is the mind and body. Through hard work, we convince our mind and body. This process of convincing our mind and body is called action. Although the soul is already convinced since the soul is infinitely intelligent and the fragmental spark of God. So when we convince our mind and body that we have worked for a fruit, we call it an action. And as so we get the fruit for that action. And when that fruit is not visible to our senses, many try to reconvince their mind and body that they didn't work harder. Even if they did. This is the foolishness.

Now let me explain what "Attitude" means. When the mind and body are convinced that the hard work is performed, the conviction is called attitude. They are the owner of that fruit, the higher this feeling is, the more is the attitude. So it's all about attitude.

So when results haven't manifested yet, the attitude must be unhindered. The attitude should be so much that one must not be extremely over-surprised by the fruit when manifested. Since he always knew that, from the start, he was the owner of that fruit. And so it's said, "It's not about the fruit, it's about the process". The knowledge about the real fruit enjoyer is mentioned to a limited extent here. So, one can always learn it in detail through the sources. So the conclusion of this chapter is, to always have a high attitude, even if the fruit hasn't been manifested. To know where is the fruit before it's manifested, one can read various sources available. This book doesn't talk about this knowledge in detail. As it will require another entire book then.

THIRTY-TWO

The Wrong Knowledge.

My friend once asked me, "What is the cruelest thing in the world". I replied," It's the spreading of wrong knowledge of the self". The knowledge of the self/soul and the God is the highest level of knowledge in the world. Comparatively, there is nothing special in the general academic knowledge. The knowledge of the self is the most important. It applies to everyone in every field of interest. Academic knowledge on the other side is about the interest of people in a particular field. But the knowledge of self is required by everyone. The educated man is not the man of a degree, the educated man is the man who has the knowledge of the self and has been applying it in his/her life. Now, why I call the teaching of wrong knowledge of self the cruelest thing in the world? Let's understand that by an example.

In the last chapter, the term "Attitude" has been defined. And in the conclusion, one was suggested to put it to the highest. Highest because, before one's fruit manifests, there are people in the society who will make you feel cursed. These people believe that the sad realities exist. One of their common statements is "We don't always get what we want". They always have a first attack on the attitude of the person. That's why it is suggested to keep the attitude at the highest. This "sad reality" group is the stupidest group on earth, they are criminals. They create a hell in the society. If you check the background of these people, they have a wrong and limited or no knowledge of the self.

These false ideas if believed, create hell. If one has the right knowledge of self, one can argue with such types of people and prove their statements wrong.

Here this out. There is no "sad reality" in this universe. No means not at all. This is one of the messages of the knowledge of self. And so, we always get what we want. All we have to do is just convince our mind and body, that we have worked for it. A strong conviction means a strong attitude. If the attitude or conviction is broken, the action doesn't count. So a high attitude is needed for an action to be considered as an action. If one thinks one didn't get fruit for his/her action, his/her attitude is broken.

Spreading the wrong knowledge of self makes one feel worthless and departs him from the devotional service. The cause of the creation of the universe. So a person, spreading the wrong knowledge of the self, is a threat. Life of such people is already miserable and they want it for others as well. The reason why their life is miserable is again because of a lack of the right knowledge of the self. They are in hell.

The knowledge of self/soul and God is the actual knowledge. The very basic and understood concept of it is that the universe has no sad realities. The knowledge of self is always nourishing. One must understand this and always have a high attitude and safeguard oneself from the wrong knowledge by the means of a rational argument backed with the right knowledge of self.

The knowledge of self and God is rational, it doesn't defy the evidence, it answers the questions of our universe, and of life. It's nourishing, it protects one from people from hell. It's the greatest knowledge available. Words would not be sufficient to explain the greatness of this knowledge.

THIRTY-THREE

The Inherited Fruit.

As it's been discussed, the attitude is nothing but the conviction that the mind and body have worked enough and the action has taken place. Now one might say, that thus the attitude is the sign that an action has taken place. But when one studies the knowledge of self. One reaches the conclusion that the most superior attitude is the attitude of action taken in the devotional service. Superior because that attitude is just not only a sign of action but also a sign of the real owner of that fruit. Here the word "Real owner" is important.

There are people on earth who inherit the fruit. And one might conclude that they received fruit without action. There is no hatred for these people.

The action of devotional service never consists of the elements of hatred, anger, envy, and jealousy. So, what is going on here? We just can't ignore this fact. This fact sounds like a "sad reality". As it shows inequality. As some people have to take action to get the fruit, while some directly get the fruit, without action. This makes people lose their faith and sometimes they get jealous of those who inherit the fruit they have to perform an action for. And of course, that's a bad thing. And the soul will never support this jealousy. But as I said, there are no sad realities in this universe. Now, here this out, "There can be no fruit without action". Saying this requires courage. Courage because how can we not agree on what we can see? There are people in this world who claim to not receive the fruit even after performing the action. This is due to their broken attitude. When the attitude breaks, one accepts the fact that he/she is not the owner of the fruit. Even if the action is performed. In the case of inheritance, the case is totally opposite. But the concept is the same. Here, the one who inherits receives the fruit even without taking any action. So who is the "real owner" of the fruit? The one who performs action for fruit or the one who inherits it? The entire game is just of an attitude. Attitude just does not make you the action taker, it also makes you the owner of that fruit. It's simply the attitude. The higher the attitude, the more real is one as the owner of that fruit. To attain the highest attitude as the owner of that fruit one is needed to perform the action in devotional service. The attitude attended then is the highest level of the attitude. This level of attitude can neither be attended by someone who inherits the fruit nor by someone who acts and gets fruit without devotional service, nor they can call themselves the real owner of that fruit.

It's been explained in the cosmic perspective how the fruit achieved by action in devotional service marks itself in the universe. This is the only real fruit. And the real action is the only action that has been conducted in the devotional service. And the attitude attained by such an action is the highest.

Such a person by soul, mind, and body is satisfied that he is the owner of that fruit.

It's always the highest attitude that makes one the real owner of the real fruit. And the real owner of the fruit always has the highest attitude. For this, one needs to act in devotional service, no matter if one already inherited the fruit or has been owning it by performing the action but not in devotional service. If the fruit doesn't convince one by soul, mind, and body that he is the owner of the fruit along with the highest attitude, the person is not the real owner of that fruit.

To achieve real ownership which comes in the form of satisfaction of soul, mind, and body along with the highest attitude, one needs to act in devotional service. So the one who inherits and the one who has to act for the same fruit both are required to perform an action under the devotional service to become the real owner of that fruit. And so nobody really inherits that fruit as the "real owner" unless the action in devotional service is performed.

Now one will ask what is the proof of being a real owner. The proof is the attitude. It's only with the highest attitude that one becomes the real owner of the fruit. People perform actions to get an attitude, and the action performed in devotional service gets the highest level of attitude. This action has to be performed by both those who inherit the fruit and those who do not inherit but are willing to perform the action to get the fruit. Thus the real action i.e. the action in devotional service is compulsory to be performed by everyone if they want to be the real owner. In this universe, "Every action is followed by the fruit of the action". And it's only the action in devotional service that gets one the highest attitude which makes one the "real owner" of the fruit.

THIRTY-FOUR

The Money.

"The money is the medium of exchange". That's what school and the dictionary teach us. We everyone use money to exchange things. And for those who have a huge amount of money, money is not the medium of exchange. Money for them is the medium to get their work done. Today the medium of exchange (money) multiplies itself, not limiting itself to just labour. This is done by investment and the investment is called an asset.

We are not talking about people having large and small amounts of money here. What is necessary to talk about is the minimum amount of money one is required to earn. Which in today's world is shockingly huge. The human need three things, food, clothing, and shelter. A human must earn enough that he is capable of buying all these three things without having a bad debt. There are two types of debt. A good debt and a bad debt. A good debt adds money to one's pocket. An example of a good debt would be taking debt for a good real estate investment. The bad debt on the other hand takes out the money out of our pocket. The bad debt doesn't allow us to do what we want. There are many examples where a person can't conduct the devotional service due to bad debt. A bad debt is not just about borrowing money, it's about engaging a person in a duty he/she doesn't like to pay that debt. The research shows that only 20% of employees are passionate about the jobs they do. And that percentage again reduces when talked about them doing their favorite work of their favorite job.

Today, the prices of houses require an average person to take a bad debt. Thus one is suggested to earn a basic amount of money, that much that he never has to take bad debt for the basic needs for his survival, food, clothing, and shelter.

With such a requirement and by looking at the prices of houses today, it can be seen that one needs to earn a lot of money by performing the action in devotional service in the field of his passion. Which is very possible.

So a bad debt is not allowing one to follow his passion. A passion is not just an interest, it's a natural attraction of a human to work in a particular field. To work is to perform an action and to perform an action under devotional service is a real action. In the real world, it is totally possible to earn enough money to avoid having a bad debt by performing the action in devotional service.

So, the observers of this universe need to follow their passion and perform action in devotional service in it. And the same action also becomes the source of his/her income. This is totally possible. All it needs is but the courage. Courage, because as an example, an average Asian family doesn't care about the passion of their kids. And it's the courage that takes for one to fight for his interest and passion and take it to the next level of achievement.

Many books on cosmic perspective ignore this important point of money. Under the true cosmic perspective not only it is mentioned, but is made clear why working in one's own interest is important.

Today's average school fails to teach about money to the kids. How to earn money is the real-world knowledge of today. The kid eats a burger for free as a student, but later in life, he has to work hours or hours to afford the same burger. This can't be called progress. Making money from passion is an art. Choosing one's own passion is as important. One can work tirelessly in his own passion and can enjoy the work for his lifetime.

So, one must earn enough to secure the basic human needs of food, clothing, and shelter without having a bad debt. And he works in his passion, where he can work tirelessly with his own will of passion. And the same becomes his/her source of income.

THIRTY-FIVE

The Relationship Between Two Observers.

Just like we are all biologically connected and were all singular back in time, we are connected to each other as a soul. A human being or observer is a soul, a soul is a fragmental spark of God. So we as a soul are the same thing. One may ignore the biological relation with one another as humans, but the relation as a soul is impossible to ignore.

And the higher intelligence of mind and infinite intelligence of soul always know this. That a bad relationship between two humans "as a soul" never happens.

A soul is an eternal fragmental spark of God, it never takes birth and never dies, it's immortal. So a mother only gives birth to a body and not to the soul. This point is important to mention because one must understand that a close and real relation is only a soul-to-soul relation. For example, a friend can be one's soul brother. Just like an action performed in devotional service is the real action. A soul-to-soul relationship as mother, father, sister, or brother is the truest. One's soul brother or soul sister is always the real and true brother or sister. And so many times it may happen that a biological brother or sister is not a soul brother or soul sister. In such cases, one's soul brother or soul-sister is the real brother or sister of the individual. The same applies for every other relationship.

A biological relative may care for a person in relation. But that relation is not always the purest. The purest love is always seen in soul-to-soul relationships.

In this chapter, the biological relationship is not demoralized. The only thing that is updated is that the soul-to-soul relationship is the truest and real relationship. We make various decisions in our life. And while doing so, whom we surround ourselves with matters the most. Among any relatives, the soul relatives are always suggested and preferred. So one must always surround himself with the soul relatives. Of course, this relationship can be a long-distance relationship as well.

This is not a book on relationships. This book talks about truth. We have been dealing with "the true cosmic perspective" knowledge. The topic of human relationships is an important topic in this discussion as it talks about the relationship between two observers of this universe. And this book is taking the observer effect in various fields of application.

THIRTY-SIX

Why Is Knowledge Of Self Is Important?

The human is given some independence on how to act. And he requires guidance to choose the right path. That's why the knowledge of the self is important. A life inside an egg has no independence, it has to hatch itself into a chick. The egg is created under the cause of life inside it. The universe is formed under the cause of observers and their actions under the devotional service. But the observers here have independence on how to behave. Under this independence, the mind and body can get wild and perform self (soul) hurting actions. Thus to bring back the observers to their original purpose of performing the devotional service, the knowledge of self is needed.

THIRTY-SEVEN

The Study Of Observers.

At this page of this book, we have come enough far to feel wrong to call a human just the "observer" of this universe. The knowledge of self is very vast. This book was just to take the observer effect at the next level. And to understand the observer and our universe and the relation between both in detail.

This knowledge must be part of school teaching. Here I am not talking about any religion or any parts in humanity. The knowledge of self never talks about religion. The basic knowledge of self is required by everyone.

This knowledge is also called the actual knowledge. Educating oneself with the knowledge of self is not a subject of interest or disinterest. It's required by everyone. Everyone must take care of their soul and ensure that the actions of mind and body do not make the soul suffer. While as soul, mind, and body one is supposed to do the devotional service. Just like the knowledge of mental health, the knowledge of self is very important. This knowledge must be taught in schools.

THIRTY-EIGHT

The Creation Of This Universe.

So, as the evidence tells us, our universe was created 13.8 billion years ago. Just like an egg is born before the life inside it, the universe has created itself before the observers, to create the observers and to fulfil the purpose of the creation of this universe. The purpose is the devotional service. Throughout the last chapters, the reader is familiar with this. In this chapter, a light is shone on the creation of this universe with the roles of observers in it.

The universe started itself from the big bang. The Big Bang was created from an infinitely dense point called singularity. In the hyperspace the space is already present. The big bang but marks the creation of spacetime. So basically, it started with the creation of time. Time is the same thing as space. The only difference is that the "time" flows.

So just at the time of Big Bang, the spacetime was created. The creation of spacetime marks the creation of the fundamental force of gravity. And after the other three fundamental forces of this universe. One of the uses of mathematics is that it helps us quantify things in this universe. The four fundamental forces of this universe have fundamental constants in their action. This fundamental constants never change their value.

Just like the egg has a right amount of nutrition present in it, required by the life inside it throughout its growth. The universe has the fixed fundamental constants that will allow the existence of observers and for the observers to do the devotional service.

After the big bang the matter started to cool down. The first element "Hydrogen" was formed. Which is also the most abundant element in our universe. After the creation of these atoms, the first star started to form. And then star systems began to form. And now to this date we have billions of star systems in every galaxy and billions of such galaxies in the entire universe. This was just done to create the observers and to support their devotional service in a single planetary system in the entire universe. This also shows the high symmetry nature of the universe.

After the formation of the sun, which is the third generation star, the saviour of the earth Jupiter was formed, and then other gas giants. Jupiter diverts many asteroids from the asteroid belt that come towards us. Many of these asteroids throughout history could have wiped out life on Earth if not been diverted by the Jupiter. After the earth was formed, in its inhabitable stage a planet "Theia" had a head-on collision with our planet. The same collision created our moon. The moon has played such an important role in the formation of life on Earth. After the earth became habitable, chemistry started creating complex bio-molecules. Then the first cell was formed, and then multicellular animals were formed. This is how life started on Earth.

THIRTY-NINE

The Evolution Of Life On The Earth.

Charles Darwin, famous for his theory of evolution has greatly explained the process of evolution. The species that cannot update themselves with the changing environment and needs of survival can't sustain. Throughout history, we have seen many species of animals going extinct. We also have witnessed the mass extinction events where a major portion of life went extinct. Such events might raise the question about our risky place in the universe. But back then the story was different. Then, the process of producing the observers was taking place. To add, these extinction events have been playing an important role in creating the observers of this universe. So like an egg is prepared with nutrition for the forecoming life inside it, the universe is pre-prepared to support the existence of observers and their devotional service. And so as the process of evolution went by, the first observer was marked in the universe when he/she realized that he/she is a soul. And from then the story of observers or humans begins.

FORTY

The Higher Dimensional Perspective.

An observer is a human being. A human being is a soul. A soul is the fragmental spark of God. So as a soul, the observer isn't limited to 4 dimensional spacetime. The world of observers works on acting and enjoying the fruits of action. Where the action taken in devotional service is the real action. And the fruit received by such an action is a real fruit. So the entire world works on performing action and enjoying its fruit. This has been discussed in detail in this book in previous chapters.

The universe we are looking at around us is just a very very small part of actual reality. When one acts for the fruit normally, he acts under the motivation of lack. So the fruit is separated by space and time in spacetime. But in reality, at higher dimensions, there is a wholeness. Where there is no spacetime separation between action and fruit of action. And the observer who realizes this and acts not under the motivation of lack is now creating from the source itself.

The study of the observers is a deep study. This is the education of "self". This book does not even talk much about the devotional service and not much about the object of the devotional service "God". This book aims to take a science enthusiast who sees a soul as an observer to the next level of understanding of the observers. And taking the observer effect to the next level of application. Application in a universe where we are not just the observers of the universe but us and our actions in devotional service as the cause of the formation of the entire universe.